Embedded

Systems

Project

Douglas R Pankretz

©2011

Embedded Systems Project

Acknowledgements

Thanks to my friend Rick Overholtzer for his help in motivating me to be myself.
And thanks to my wife Vickie for allowing me to do what I wanted to do.
This is almost a work of fiction or fantasy if you will.

In other words: Not to be taken too seriously.

Disclaimer

Note: The title on the cover is meant to convey my understanding of
the work of God behind the scenes of life itself: God does in fact
get through to some people. The Holy Bible is evidence of this.
The Word of God is embedded within Life itself.

As for my self, there are times that I believe it is God that compels me
to do the things I do. Some things I can see for my self that I knew
what *"I"* was doing. While being creative, though, I believe it comes
from God. "Whatever good you find to do with your life, do it with
all you heart". Besides that, I say that reading the word of God helps
me to keep on speaking terms with Him.

Even so, I cannot tell how what I have done will have any influence on
bringing others to Christ. Unless they are some kind of intellectual
like me; with an interest in Science. There have been many times when
I was reading a science book and I did not agree with what they say.
I felt like God was showing me His way to see reality.

2010 : First Rendition – E-mailings to friends

2011 : Second Rendition – Self Publication

ISBN: 13:978-1460910221

Credits: ClipArt from Microsoft™ Word Online

Reality Check

Dedicated to

Stephen W. Hawking

Ecological Validity,

Is the test for integrity.

It's the factors we consider;

To make our theories better,

To include the obvious evidence,

Or what we think has relevance?

For the models to the mind of man

Do help him to understand;

The rules first exception is sure,

... At the rules inception.

Embedded Systems Project

Embedded Systems Project

<u>Various Contents</u>

0. Sunny Talk-Mun
1. Einstein's Challenge
2. Translational Discrepancies
3. Energy and Light
4. Frame Dragging Theory
5. God in the Electron
6. Area of Circle in a Square
7. Fractal Features of Gravity
8. Accurate Models of Reality
9. Three Dimensions of Time
10. Four Dimensional Brain
11. Impossible Dreams

Appendix

Embedded Systems Project

Sunny Talk Mun

There is a new conversational companion on the market that will give you just hours of dialog at you own level of interest. You can just talk to it and it will make a proper response for you.

First you start by telling it your IQ: just to give it a clue as to where to start. For example you can say 180 and it will carry on about Nuclear Science and keep you up to date with the most accurate Model of Reality. Current research going on at the Hadron Collider; and discoveries in astrophysics.

Tell it 150 and you'll be getting into Politics and even become an activist. Before you know it you'll be running for office, so that you can help change things around here.

Try a 120 and you will be steeped in Religion and Economics for as long as you have an interest. Reasons for their connections will be clarified (supply and demand).

At 100 you are more down to Earth with talk about your favorite TV shows. It can keep you posted with brief reviews of ones you missed and remind you of ones to watch for.

Around 85 you are limited to soaps and gossip about your own friends, sorry to say. Don't ask me how it could know what they have been doing (Face book, Twitter, other social networks).

Down wind of 65 it'll comeback with a raspy accent "This here's Rap Core. What's your handle? You got a copy? Come back now." This thing will appear to have some personality.

If you want to get testy; try and IQ below 50. You get nothing but questions and hostile comments like: "See what you made me do!", "Just who do you think you are?", "Get down off your high horse and help me out."

Einstein's Challenge

One thing that got me going is this: "No scientific experiment can be done to prove whether you are moving or not – when you are within a closed room (no windows or doors)." That is something that Albert Einstein is said to have stated.

Well, I think differently. One can use a pendulum to find out if one is moving or not. There is something about the rotation of the earth that will have some influence on the precession of the swing of the pendulum. This was learned by Leon Foucault in 1851. He was the first to <u>prove</u> that the earth rotates. Truly I am surprised that Einstein did not understand that one could use a pendulum in his closed room to determine whether that one was moving or not.

Leon Foucault found a simple equation to match the precession of the pendulum. This is dependent on the latitude 'θ' of the pendulum: $t = 24/\text{sine}(\theta)$. So the time of precession at 34° is about 43 hours. If you are located at the pole (north or south) the sine (90°) is 1.0 so the time (t) is equal to 24. If you are located at the equator the sine (0°) = 0.0 so the time (t) is ∞! In other words it does not rotate from the plane of the swinging pendulum.

What we have is a way of finding out whether we are in fact moving. Consider motion along the longitude: The rate of the precession will keep changing. As we move toward the equator it will slow down. Toward the pole it will speed up. If we are not moving it will stay constant.

You may think that if we are moving along the latitude then motion might not be detectable. But, maybe we can start two pendulums at right angles to each other – some distance apart, of course.

This is one experiment that may have to be carried out. I can only believe that some variation will be obvious. Even if you are moving at a constant velocity centripetal forces will be evident in the swing of the pendulums. Not knowing which way is north, it will become apparent in the distance from center that the end of each swing makes; of this I am sure.

Translational Discrepancies

Scene 1: View toward the window. Professor Now is at a table working on calculations for equations. Table is covered with books and tools – several puzzle games are visible. Pendulum swings from ceiling hook.

Professor Now
"According to these calculations we can figure out what our latitude is, by measuring the time it takes for the pendulum to presses' around the floor."

Scene 2: Camera is behind him with door visible in background .Door opens. Video camera appears and Professor looks up to see who is entering. Uncle Rick enters with microphone at hand:

Uncle Rick
"In here we have our latest project in development. May I introduce Thee Professor Now?"

Scene 3: Camera is behind Uncle Rick, with Professor Now in view over Uncle Rick's shoulder.

Professor Now
"Ah yes. Yes come in. We have evidence here that the Earth rotates. You may see it for your self if you have the time to watch this pendulum swing."

Uncle Rick
"Isn't that something that was proved 150 years ago?"

Scene 4: Camera shows both in view. Pendulum is in the background, swinging back and forth.

Professor Now
"Indeed it was, by Leon Foucault in1851, in the Pantheon in Paris."

Uncle Rick
"But the Bible states as much many years before that, right."

Professor Now
"Yes. If you like I will read it for you, from Job 38:12"

Embedded Systems Project

Scene 5: Camera is behind Uncle Rick focused on Professor Now.

Uncle Rick
"My pleasure, please continue."

Professor Now
"This chapter starts with the Lord answering Job from the whirlwind in verse 1, but in verse 12 it says 'Have you ever commanded the morning to appear and caused the dawn to rise in the east?' "

Uncle Rick
" '… to rise in the east … ' I take it to be what you are referring to?"

Scene 6: Camera up close on Professor Now.

Professor Now
"Isn't that obvious? Nowadays we know the earth rotates and the sun will 'rise' in the east."

Uncle Rick
"Earlier you were telling me about a recent discovery you felt most ambitious about."

Professor Now
"First you want to know how this is relevant, right?"

Uncle Rick
"Yes. Is it related to the project you are working on?"

Professor Now
"I think so, maybe off on a tangent. I should hope it is not to be going perpendicular!!"

Uncle Rick
"Do you have a bible verse to begin with?"

Professor Now
"Isaiah 55: 8-9 'My thoughts are completely different from yours,' says the LORD. 'And my ways are far beyond anything you could imagine. For just as

the heavens are higher than the earth, so are my ways higher than your ways and my thoughts higher than your thoughts.' End quote."

Uncle Rick

"How is that relevant?"

Professor Now

"We have something here, stated by God in the Bible, which only recently was stated by Scientist as "The map is not the territory."

Uncle Rick

"I may have heard that one."

Professor Now:

"'The Model is not The Realty' is what I had to figure out for my self while at College. That statement means the same thing, to me, in my own terms. One other way to help us clarify the concept is that blueprints are not the buildings they represent.

Uncle Rick

"Obviously! I can agree with you on that."

Professor Now

"Something gets lost in the translation from one language to another – the culture. The words in the original text have meanings that don't come across entirely."

Uncle Rick

"This is an example of what you mean by 'coming to terms' with things that you've read?"

Professor Now

"There are times when I have to rewrite what I am reading, on the fly – so to speak. "

Uncle Rick

"How does that relate to what you quote from the Bible?"

Professor Now

Embedded Systems Project

"Our thoughts, our ideas, our theories are all abstractions. They are models of our observations. Which reminds me of a poem, several actually. [Warning, Danger, Beware, Caution]."

Uncle Rick
"Can you share one of them just now?"

Professor Now
"If you don't mind I will just rattle them off, one after the other."

Warning
Some of our thinking, Is truly deceiving, Us into believing;
What we're conceiving, Is totally revealing, About our own reeling!

Danger
The major attraction, To partial abstractions, Is to model reactions,
In search of infractions, To such a distraction, It puts us in traction.

Beware
The implications, Of limitations, To extrapolations, Of imagination
Have undulations, Of indignation, For the duration, Of application.

Caution
Concepts of Reality, Which convey a Mentality, Of Truth or Morality,
Shall endure Mortality, In subdued Frugality, Through songs of Duality.

Uncle Rick
"Wow! I mean whoa, that will be enough for now. I'll have to get a copy of that from you to read later. It really is too much all at once. Maybe we can share them with our audience."

Professor Now
"I'll leave you just one more before we part company, Vaccine."

Vaccine
Purity of truth is for the intellectual - Use caution with those of not intentional
Some of them say it may be detrimental - Others insist that it will be beneficial
Either of which it is influential - Not as merely, one existential.

Energy and Light

Today we have a talk about energy. First we will read a few verses from the bible:

Genesis Chapter 1: verses 1-5

"In the beginning God created the heaven and the earth. And the earth was without form, and void and darkness was upon the face of the deep. And the Spirit of God moved upon the face of the waters. And God said, "Let there be light" and there was light. And God saw the light, that it was good: and God divided the light from the darkness. And God called the light, Day and the darkness he called Night: and the evening and the morning was the first day."

Our point of interest is '*light*', what we now speak of as <u>Energy</u>! Isaac Newton is credited with finding that light splits into a spectrum of colors. What we now call frequency or wave lengths. There are several things to learn about energy. One is that the frequency spectrum spans from a very low energy (ULF): long wave length (a million times longer), used by submarines communications – to the very high energy (X-Rays): short wave lengths (a million times shorter), used by medical personal.

The second thing is that the speed of the energy varies with the frequency; because of the intensity of the energy. A long wave length *means* a low energy thus a slower wave front. Same with a short wave length *means* a higher energy thus a faster wave front.

So what if the speed of light is constant?
I am talking about <u>energy</u> **beyond** the spectrum of *light!*

What I have to say about this has to do with what Einstein discovered: $E = mc^2$ One look at that formula and I realized a poem:

Quantum Ocean

Matter and Energy, with Information;
Appear at Once to Bind Observation!

It was with some hesitation that I realized that I was looking *at* Information! It was embedded within material elements and it took some *energy* for the information to get from the paper into my own brains/mind even.

Like: Matter and Energy are two sides of the same coin; Information could be on the edge of the coin – this Reality. We, the Observers, are the 'coin' itself, a **spiritual** being.

Where do we go from here? Just this: we have come to terms with understanding God's own creation. I can say Praise the Lord for a spark of intelligence. We can take what other people have written, reflect on it with the help of God's word, and discover our own understanding of various meanings in life. Here we can say the speed of 'light' varies with frequency (light/energy).

One day I heard some 'scientist' talk about the void of space within our own bodies. "Between the nuclei of atoms space is empty." Ha! I say, Energy exists in the 'empty' space, giving my body its temperature. Behold I have a new definition for Temperature – it is the Density of Energy.

Wow, suddenly you can understand why the temperature of the sun's corona is hotter than its surface. The density of energy varies with the amount of matter that permeates that volume of space.

Next thing you know the temperature of a Neutron star is absolute zero. Because: Energy does not exist between the particles of matter (neutrons). Could it be that the nucleus of elements such as; carbon, iron, lead, etc. have the same condition? That could be the one thing that is holding the nucleus together. "They" call it the strong force. I say they are frozen with a temperature of absolute zero.

Now back to the Word of God

Isaiah 55: 8-9 'My thoughts are completely different from yours,' says the LORD. 'And my ways are far beyond anything you could imagine. For just as the heavens are higher than the earth, so are my ways higher than your ways and my thoughts higher than your thoughts.' End quote.

Stephen Hawkins is said to believe he would hope to read the very mind of God – something to that effect on the last page of his book: "Brief History of Time"

I would like to end with a quote from Einstein: "The most incomprehensible thing about the universe is that the universe is comprehensible."

Truly, all I can understand is our Models of 'reality', our theories and beliefs. Steeped in language as much as they are, I can only test them for accuracy, or accept them as plausible.

Frame Dragging Theory

Frame Dragging Theory – "if any of you lacks wisdom, ask of God who gives liberally"

I began to wonder about the 'Jet Stream' in our atmosphere – why does it move in the same direction as the rotation of the Earth. How could it if the rotation of a mass is dragging space itself? Talk with God and you will find that the 'empty' space is filled with energy. That field of energy is in motion. The entire universe 'rotates' and thus adds to the expansion of the universe. This Rotation has a fractal feature to it also, at so many different levels.

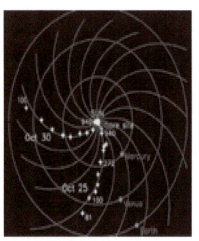

Scientists have it all wrong ... I must say that the energy field of space is that which is rotating and dragging mass in its wake. Take a look at the pictures from the space craft Magellan at the south pole of the sun. The spiral of particle emissions is going in the direction in which the planets are rotating. From the South Pole it is a clockwise direction.

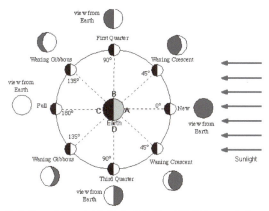

The Sun-Moon angle is the angle defined by Sun->Earth->Moon with Earth (where *you* are) as the angle vertex. As the Sun-Moon angle increases we see more of the sunlit part of the Moon. Note that if this drawing were to scale, then the Moon would be half this size and its orbit would be about 22 **times** larger in diameter and the Sun would be about 389 **times** farther away than the Moon!

So the same it is with the rotation of our moon around the earth. It is being dragged around by the energy field of space. Take a closer look at the surface of the earth and you will see the jet stream flowing in the same direction as the moon is rotating. Thus it just seems logical that there is an 'ether' of energy in space that maintains the rotation of the sun and rotation of the planets. Talk about Frame Dragging: it is the swirl pool field of energy that is dragging the mass of planets around the sun and the moons around their planets!

Within our atmosphere we have *high* and *low* pressure zones that seem to be gravitational vortices. That is to say that the rotation of each zone is due to a dynamic gravitational field. It maintains the rotation of the earth <u>and</u> the flow of jet streams in our atmosphere which helps in the dynamics of the high and low pressure zones. Yet, it scares me to think that this is the truth about reality.

A virtual Hand of God maintains the orbits of planets and moons; rotations of suns and planets; and the flow of weather changes.

God in the Electron

I found God in the Electron. Here is how.

Some time ago I was reading about Scientists and their statements about the electron as a point particle within a cloud. Like maybe there is no electron per se, just the cloud. I wondered if the electron might be the point of infinity, even if it could be the Infinite its Self. Truly if it is God, then the electron cloud does not get bombarded by energy wave fronts of photons – it generates them! This is where God controls everything in the universe. We are within God and God is within us …

This could be used as the foundational belief for a holographic universe!

I spent 30 years reading about nuclear science; trying to keep up with the latest developments.

Some of the scientists tell about their discoveries of what things are made of, e.g. Atoms are made of Protons, Neutrons and electrons. Then what are those are made of; quarks, clouds, strings … And then what? They would like to know what are the subatomic particles made of? Nothing?!

Maybe the jokes on me or is the joke on them? What they have is basic structures in abstractions: words with definitions. Nothing more…

If you think about it; can 'science' possible know every thing about Reality? When 'They' get to the bottom of everything they find there is 'nothing'. Could it be that they have found God, but they cannot see Him because He is invisible!

Just as an aside I have the idea that nuclei do not lose mass when they combine. Protons and neutrons have a fog of energy that surrounds them when they are separate from other particles. When they combine, the particles are like soap bubbles – they cling to each other and the 'fog' between them is displaced and dissipates. This is considered, by some, as a loss of mass (one percent even).

Truth of the matter is that the standard number for the mass of these particles does not give an account for the presence of energy that surrounds them. You have to understand this: where matter exists energy does not exist; it has been displaced. The field of energy exerts a force in reaction to this displacement. Just like a sponge being displaced by a marble. The sponge will stretch around, parallel to the surface of the marble and it will compress vertically (perpendicular to the surface).

Area of Circle; in a Square

Use a straight edge and compasses to find the area of a square equal to the area of a given circle. They said that it could not be done (seventh/eighth grade math). You may try this yourself and check the accuracy of your results using a metric ruler if you wish.

Well, it seems like 'everybody' has done it now. So, I have finally put my version of Squaring a Circle together in PowerPoint. I did the math on it back in 2003. It was a lot of work checking the angles and values using trigonometry. My result came close to the value of PI with an accuracy of 8 decimal places. Using pencil on paper the accuracy was only three decimal places.

I got started with a picture of a square with circle inscribed and a sketch of the square with the *hopeful* value of PI for its area (at a slight angle). Then by using the Pythagorean Theorem and the Quadratic Equation; I found the value for x given that: $(2 - x)^2 + x^2 = 4 - 4x + 2x^2 = PI$ (x = .24448936).[x is the value from the corner of the outer square up to the corner of the inner square contact.]

Here is where I 'lean not unto my own understanding' and listen to God for guidance. I got the idea to bisect the diagonal segment between the circle and corner of the outer square. After that, well you may see in the appendix. Anyway, I was just totally amazed how accurate the result was: 8 decimal places.

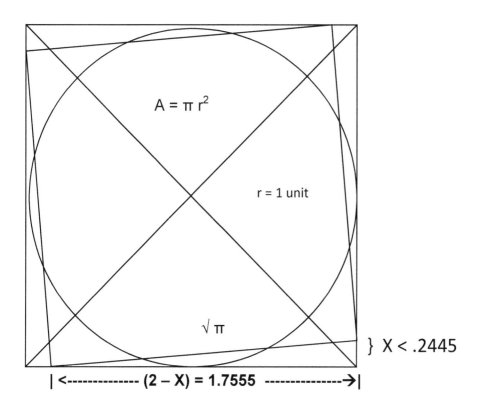

$A = \pi r^2$

r = 1 unit

$\sqrt{\pi}$

} X < .2445

|<--------------- (2 – X) = 1.7555 --------------->|

$(2-x)^2 + x^2 = 4 - 4x + 2x^2 = (\sqrt{\pi})^2 = \pi$

Fractal Features of Gravity

<u>Einstein tried to tell us about the curvature of space.</u>

I tell you, "Space is twisted with swirl pool fractals."
- Each level of the swirl pool is a smaller copy
  of the larger image (fractal feature requirement).

The Milky Way Galaxy is "Level one" of a stable vortex.

- Comparable to the *trunk* of a tree (e.g. Dogwood).

Our Solar System is the next level, also a stable swirl pool.

- Compared to the *limbs* from the trunk of a tree.

Next level is the planets (w/o moons), in a fairly stable vortex.

- Compares to the *branches* from a limb of a tree.

At last, down to Earth, on the surface of a planet we have
distortions in the gravity field, hurricanes and tornadoes.

- Comparison to flowers and leaves on a branch.
  These are not only quite different from each other,
  they are different from the rest of the tree.

Now then back out beyond the galaxy is the root of all creation.

- Comparatively all galaxies have the same root system.
  Indeed they are all connected and yet appear separate.

Such as it is then; who can say if the weather is "in effect"
Frame Dragging – to the extreme (high and low vortexes).

- Wherever matter exists space/energy does not exist/coexist – it is
  displaced.
  Space/energy exerts a force in reaction to this displacement as
  gravitational field.

-

Embedded Systems Project

- The energy of space can be measured in three different ways: Three **Dimensions;** Gravitational, Magnetic, and Temperature.
- Like matter has three dimensions also: Length, Width, Height.

Therefore space is not empty; it is filled with energies of different densities. I am sure and yet not certain the expansion of the Universe is along the orbits of the planets away from the sun.

- We have a separation of the heavy elements from the light elements, planets from the stars.
- We are on the surface of one boundary of space/energy and the other boundary is on the surface of the Sun.

Accurate Models of Reality

Philosophies' Tentative Assertions
1. Models are not what they Actually Represent.
2. Models do have a Reality unto themselves.
3. Fantasy is a Model with no Actual in Reality
4. There is more to Reality than anyone can Fantasize.
5. Each of us has our own Fantasies of Reality.

Conceptual Integrity Axioms
1. Focus of Attention (internal/external)
2. Personal Perspective (opinion/factual)
3. Motive/Intention (motivate/informative)
4. Individual Experience (learned/taught)

Nature of Reality
1. The nature of Matter is one of Order it is Stable; it is Static.
2. The nature of Energy is Chaos; it is dynamic – Change.
3. The nature of Information is one of Magic, it is influential.

Quantum Ocean
Matter and Energy with Information;
Appear at Once to Bind Observation

Quantum Mechanics
They Make a Noise and Strike a Pose;
To Emulate the Quantum Rate's -
    Intrinsic Puristic Statistic!
With Thinker Toys for Super Ploys;
To Speculate and Postulate -
    Simplistic Heuristic Logistics!
We'll Take your Mind through Space and Time;
To fabricate and Replicate -
    Stylistic Dualistic Dialectics!

Three Dimensions of Time

There was a time when I was talking with a friend about time. I asked him to consider: "What if our past and future are at odds with each other?" I stretch my arms out to show linear. "What if Time is not linear like this?"

Then I move my left arm out front. "What if they are at an angle?" What if they are orthogonal like the diagram below?

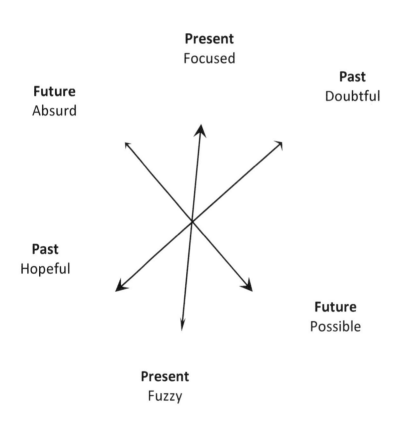

Technical Difficulty

What I learned from this diagram is about our attitudes; how it is that they seem to change from positive to negative when we are talking with each other. They have some oscillations as we speak – with or without being conscious of it though; I have to wonder.

Some people may have grand visions of the Future, when I listen to them I consider some things to be *possible* – I feel good about them. I'd like to hear more of the story; other visions I think are *absurd* – so I feel like discouraging them as they speak. There goes my attitude, as if I had a choice.

Other people can tell me of the wild times they have had in the Past and of course if they are what I think is *doubtful* then I frown at those stories I hear. When I believe it then I am joyful at believing a *probable* truth. Again, did I have any choice in my change of attitude?

Then we have a few people who like to be telling me what we are doing for the moment, at Present, so I am *focused* when I understand; otherwise I may be confused and my attitude is negative with *fuzzy* thinking.

So, there you have it. Does any one have a choice in their attitude when they react to the stories that others have to speak of? Our three dimensions of time co-exist along with our selves.

Then again, what about the goings on with internal dialog? For those who talk to themselves.

There may very well be a Fourth Dimension of Imaginary Time wherein people lie/exaggerate?

This is something that I would like to delve into just for a moment. Consider my own stories.

Have you noticed any changes in your feelings about what you have read? Is it encouraging, like what I have written is truly worth consideration? Is anything making you feel that I am stretching the truth a little? Do you have some response to what you read? Does the writing speak for itself? Are there some unanswered questions?

Example: Blind Date

What we have is a way to explain the different attitudes that people have about their various perspectives regarding Time: Past, Present and Future. Our attitudes seem to change on the spur of the moment. Clearly what comes into play here is the positive and negative attitudes people can have when they consider various ideas that make reference to events in the Past, Present or Future in their own lives; in the course of conversation.

Imagine two people talking about plans for a trip: whether to go to the beach or up to the mountains. Judy presents her possible adventures for a climb in the mountains for a view above the valley. She is all enthusiastic – totally positive attitude; eager to go. Bill considers her plan as absurd and presents his idea of wind surfing for the adventure of speed. He in turn is very much ready to go.

Well of course, she considers his idea absurd also. So here we have two people with conflicting attitudes about each others hobby just for starters. Each in their own frame of mind being positive about what they want to do, in the Future. And also each in their turn, being negative in response to the others Future possibility.

The next thing they do is talk about their past. Each of them presents their stories of what they have done with their hobbies, as an attempt to encourage each other to go along anyway. As they tell their stories you can imagine that each has some doubts about the truth of the matter being spoken. They do have a tendency to exaggerate. The thing of it is that neither of them has experience being involved with other people who have the different hobby as they do. Bill was raised on the beach and Judy grew up in the mountains.

Now for the Present consideration: Their meeting began somewhat focused in their agreement to make plans doing something together – both were in a positive mood with their understanding of being open to suggestion; allowing each other to have their say.

In the end they were both kind of negative, attitude was fuzzy, about having to rent equipment for them selves, to join in the fun of doing things together. So to get focused they chose to go white water rafting and split the cost!

Four Dimensional Brains

String of Magic; Books!

Allow me to explain what I call Non-Linear Geometry.

First off, a line of one dimension is called 1-D – essentially straight; no bends, no curves, no twists even; unlimited – straight and narrow, to the max; Mathematically.

Second place, 2-D is a plane of two dimensions – essentially flat, no warp, no curls, and no folds even; infinite – figuratively, of course.

The last thing we need is a cube in 3-D, with all of six sides flat (2-D), 12 edges straight (1-D). No distortions, no inflation no ripples either.

Now for the First Magic Trick – I take the line of 1-D and bend it to fit onto the plane of 2-D such that points of the line are like a Map to points on the plane (see below). I used a short line segment and a small square plane. This line is now said to be nonlinear, it is more like 1.2 dimensional. Example of this is a tape measure: it coils into a box, no longer a straight line. Mathematically; the values on the tape are still linear. As an abstraction they 'map' onto a 2-D surface. [1-D line 'points' to a 2-D plane]

The Second Magic Trick is to take a two dimensional plane and fold it up inside of a cube. So then, now the plane is no longer 2-D it is more like 2.3-D. Here again the points on the plane match points within some parts of the cube. Origami is a good example for folding 2-D into 3-D.

For the Third Magic Trick you take a Rubik's Cube™ and twist it around until you have a mix of colors on each side. It then is no longer 3-D. The points, lines and planes of the cube then match up with/in the Hyper-Space of 4-D (as if there really are 4-D objects beyond imagination – spiritual?).

Here is where I get off with saying "Our brains have the four dimensional geometric structure (tangled brain cell synapse) that gives rise to the illusion of time." And besides that, I say our writing is a non-linear "String of Magic". Each word is a sensible piece of a map fragment; pieced together from letters, or 'strings', of the Alphabet. A sentence maps onto a concept. Making Sense out of Non-Sense.

Embedded Systems Project

Think about it, write about it, then rewrite and rewrite …
Rewrite Strings unto Magic Books!

The Magic of a Secret is the Secret of Magic

Impossible Dreams

Ultrasonic Fire Extinguisher
– Three parabolic fields of ultrasound at right angles to each other

Cyclotronic Fusion Machine
– Cyclotron used in reverse design, with funnel collector

Ezekiel's Vortex
– Space travel without any propellants – energy compression

Terabyte Solar Power
– Terabyte hard drive technology used for solar power collector

The Appendix

From the Finish with a Start

How to Square a Circle with Compass and Straight Edge

$$A = \pi r^2 \qquad r = 1$$

Design by Douglas Pankretz
2010 Copyright

$$(2-x)^2 + x^2 = 4 - 4x + 2x^2 = (\sqrt{\pi})^2 = \pi$$

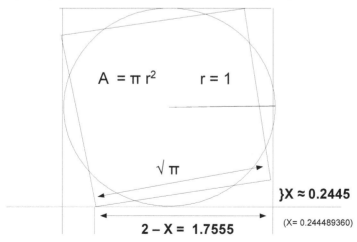

$$A = \pi r^2 \qquad r = 1$$

$$\sqrt{\pi}$$

$$\}X \approx 0.2445$$

$$2 - X = 1.7555 \qquad (X = 0.244489360)$$

Embedded Systems Project

Draw the base of a Square

Establish End Points for Sides

Draw the Sides of Square

Draw Arcs to the Top of Square

Draw the top line of Square

Locate center for Circle

Draw Circle in Square

Bisect all short segments

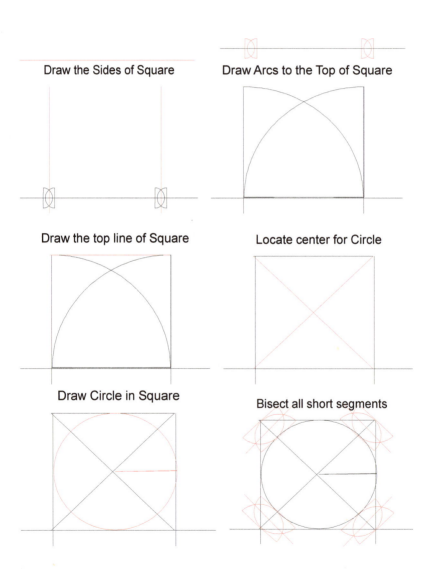

Embedded Systems Project

Remove arcs of bisection

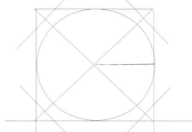

Perpendiculars for Magic points

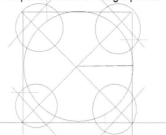

Pause for clean up again

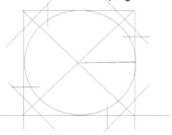

Intersect bisector to Magic points

Remove bisectors & perpendiculars

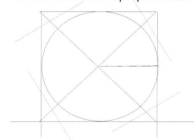

Draw the Square of the Circle

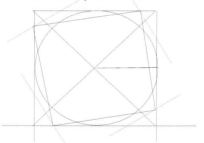

Inner Square is same area as Circle

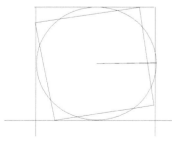

$(2\text{-}x)^2 + x^2 = 4 - 4x + 2x^2 = (\sqrt{\pi})^2 = \pi$

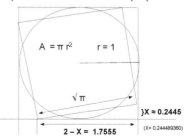

$A = \pi r^2$ $r = 1$

$\sqrt{\pi}$

}X ≈ 0.2445

(X= 0.244489360)

2 − X = 1.7555

Embedded Systems Project

www.ingramcontent.com/pod-product-compliance
Lightning Source LLC
Chambersburg PA
CBHW041147050326
40689CB00001B/524